This book belongs to

____

For Phoenix,

Be curious and kind

Written by Andrew Sario
Illustrations by Mari Acierda

Published by Engineering IRL 2024
engineeringinreallife.com

ISBN 978-1-7637573-0-1

THE ENGINEER'S DUCK

Book 1 of 8 in The Engineer's Duck series:

Engineering Picture Book Stories For Curious Young Minds

In a workshop bright,
where gadgets gleam

Sits a clever engineer named Lon Lon

with a curious dream

He had a trusty sidekick, a duck by his side

Together they ventured on an engineering ride
WORLD OF ENGINEERING

The engineer's duck,
with feathers so yellow

Helped

Lon Lon

solve problems,

what a

wonderful

fellow!

"What's on the moon?",
asked a curious person they met

So they used spaceship blueprints,
to launch a three stage rocket

They built a suspension bridge,

sturdy and strong!

Helping people cross rivers,

all day long

They designed a robot,

with gears and cogs

It could dance and sing, even chase dogs!

They built websites and apps

by hacking some code

Connecting family and friends
across the globe

They built a drone,

precise and swift

To lend a helping hand

and deliver a gift

They constructed a dam,
blocking the flow

Harnessing water to make electricity glow

They engineered a car, fast and sleek

Driving tunnels and roads with a joyful squeak

They built skyscrapers that

stand strong on a windy day

Touching the sky is no problem, It's designed to sway!!

They invented smart things,
to face problems gigantic,

From daily feats to
exploring the deep
zap, done like magic!

Young Engineer

Lon Lon and his

duck named TED-e

MECHANICAL
DIGITAL
ELECTRICAL
ROBOTICS
Now look to
help others,
listening curiously
HYDRO
AERONAUTICAL

In a world of

innovation,

they make their

way through

The engineer and his duck,

making dreams come true

The Engineer's Duck
Point-and-follow reading

Did you follow along?

Point at each word

you read out loud!

THE ENGINEER'S DUCK
World of Engineering
Book 1
Andrew Sario
Mari Acierda

Book 2

Book 3

Book 4

Book 5

Book 6

Book 7

Book 8

THE ENGINEER'S DUCK
Series

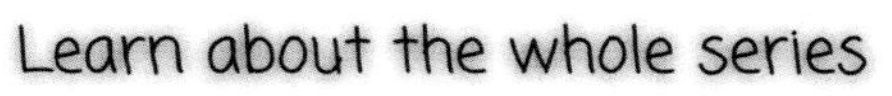
Learn about the whole series

If you enjoyed learning about science and engineering you may be ready for the TED-e academy!

You get access to science and engineering activities using just household items and you can share your achievements with the growing community.

You can also get a Certificate of Completion for your hard work!

Just scan this code to continue your engineering journey

Made in the USA
Middletown, DE
15 December 2024